Josiah Coates

Concise

Geometry

To Sole

Contents

Chapter 1 Lines and Points..1

Chapter 2 Planes 4

Chapter 3 Angles 7

Chapter 4 Planar Lines 12

Chapter 5 Triangles 16

Chapter 6 Polygons 29

Chapter 7 Quadrilaterals..36

Chapter 8 Similarity..39

Chapter 9 Right Triangles 42

Chapter 10 Trigonometry..48

Chapter 11 Circles 60

Chapter 12 Perimeter and Area 71

Chapter 13 Volume and Surface Area..82

Chapter 14 Transformations 93

Questions or Comments?..100

Preface

A typical high school geometry textbook can be over 600 pages long, full of complex theorems and postulates. Working as an engineer for a major international shipping company, I have spent my entire career explaining complex subjects to senior (and highly paid) executives. If I used a "math textbook" approach to communicating with executives, I would have lost my job a long time ago.

Executives, it is believed, are very busy – and they need quick and concise explanations to make quick and well-informed decisions. But are high school students any different? Do they have endless time to read 600 pages of postulates and theorems that have no application to modern life?

To be fair, geometry – along with algebra, trigonometry and calculus – does have some application to the modern workforce. But it is the general concepts of these subjects that are applicable. The onerous theorems, postulates and definitions – lengthy explanations using incomprehensible terminology are, in my opinion, useless. This approach to math should only be taught to students (college students) that have decided to become mathematicians. And there are quite a small number of mathematicians in the world.

When I prepare papers and presentations on the most complex engineering topics, topics which are often cutting edge and cannot be easily researched on the internet, I continually revise and revise until the presentation is short, concise and understandable by almost anyone. It would seem only logical that a high school math textbook should be prepared the same way.

The subject matter in this book covers an entire geometry course and can be mastered after about 20 hours of study. It should take an additional 10 hours to master the practice problems at the end of each chapter. This book is ideal for summer study, catch up, tutoring or self-tutoring.

For any questions or comments, I can be reached at info@concisetextbooks.com. I will respond to any requests for clarification, as I am interested in revising the text in any area where it may be hard to understand.

Sincerely,

Josiah Coates

Chapter 1 Lines and Points

A single point is represented by a dot, such as point A below:

A line is represented by arrows on each end (and in theory the line goes on for infinity), such as line AB below:

A line segment is represented by points on each end, such as line segment AB below:

A ray is represented by an arrow on one side (indicating the line goes on for infinity in that direction) and a point on the other side, such as ray AB below:

In the below figure line segment AB and CD are equal (AB = CD), as indicated by the strike mark between each line segment:

A   B   C   D

In the below picture, the length is = 4+6 = 10

Midpoint = 10/2 = 5

The Midpoint is at point F, therefore AF = FK

-4 -3 -2 -1 0 1 2 3 4 5 6

A B C D E F G H I J K

In the below figure, A, B and C are collinear (all on the same line), but A, B, C and D (or A, B, D or A, C, D) are not collinear:

A •

B •   • D

C •

Practice Problems

1. Match up the image with its line type:

1.  A — B   a. Line AB

2. 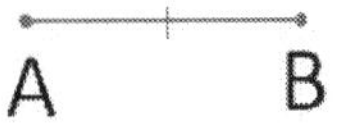 A — B   b. Ray AB

3.  A — B   c. Line Segment AB

4. A ·   d. Point A

Answer: 1. B, 2. C, 3. A, 4. D

2. Determine the midpoint of the line below.

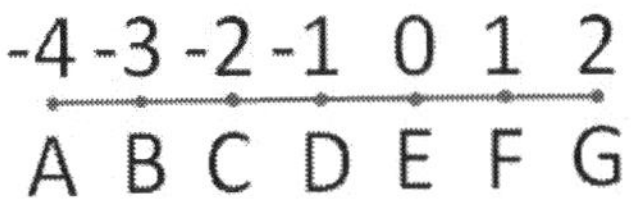

$$-4 -3 -2 -1 \ 0 \ 1 \ 2$$
$$A \ B \ C \ D \ E \ F \ G$$

Answer:

The line is 6 segments long:

$$6/2 = 3$$

$$-4 + 3 = -1$$

The midpoint is at -1

Chapter 2 Planes

Planes have two dimensions: length and width. Below are some examples of planes:

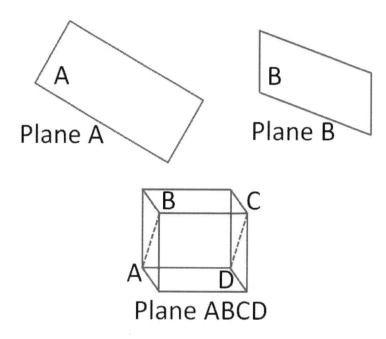

Coplanar refers to two or more points located in a single plane. Points A, B, C, D are all coplanar (located in plane ABCD).

In the figure below, three lines (b, c, d) are in the same plane whereas *line a* lies on a different plane. *Lines b* and *d* are parallel, and *line c* intersects *line d*. *Line a* misses *lines b, c, d*.

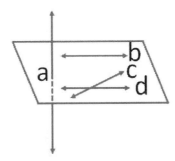

In the below figure, *line b* bisects *line a*:

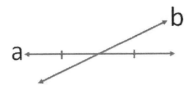

Practice Problems

1. Which set of points make up a plane (ie, all points must lie in the same plane):

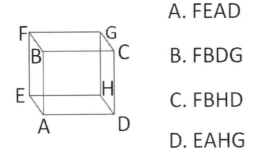

A. FEAD

B. FBDG

C. FBHD

D. EAHG

Answer: FBHD are the only points that lie together in a single plane

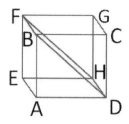

2. In the figure below, which lines are in the same plane?

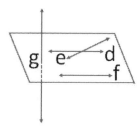

Answer: lines d, e and f all lie in the same plane. Line g lies in a different plane.

3. Which of the following lines are bisected at their midpoint?

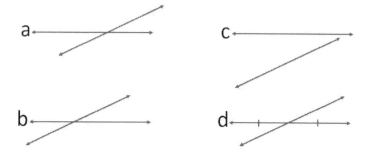

Answer: Line d is bisected at its midpoint, the point of intersection creates two congruent line segments.

Chapter 3 Angles

B is vertex of the below angle. We can refer to the angle as ∠B or ∠ABC.

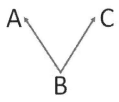

In the figure below ∠B is acute because it is less than 90°, ∠E is a right angle because = 90° and ∠H is obtuse because it is greater than 90°.

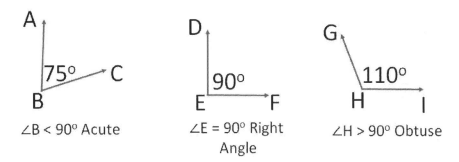

∠B < 90° Acute   ∠E = 90° Right   ∠H > 90° Obtuse
                                      Angle

The below angles are congruent (congruent means equal) because they have the same degree measurement:

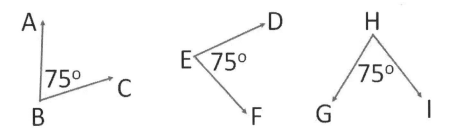

Here: ∠B ≅ ∠E ≅ ∠H (∠B is congruent to ∠E is congruent to ∠H)

Congruent angles can also be indicated by using a small arc near the vertex:

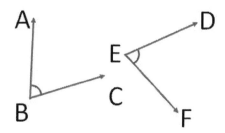

Different congruent angles can be indicated by using multiple arcs to differentiate from each other:

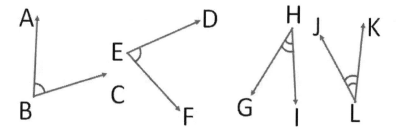

Here: ∠B ≅ ∠E and ∠H ≅ ∠L

The opposite angles of intersecting lines are always congruent. Notice how the sums of the angles always add up to 360°:

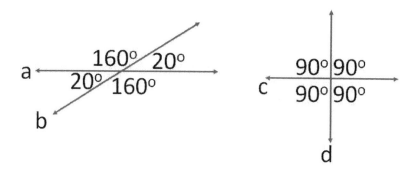

Complementary angles are angles that add up to 90°, and supplementary angles are angles that add up to 180°.

Complementary (90°):

Supplementary (180°):

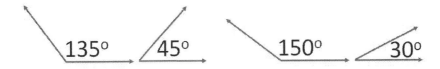

Ray D bisects the angle below. To "bisect" means that the ray divides the angle equally in half. This means that ∠ABD = ∠DBC (NOTE: in this case it is not correct to refer ∠B because there are two angles at the vertex B):

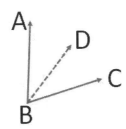

Practice Problems

1. Which angle below is ∠ABC?

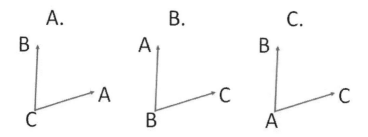

Answer: B is the correct angle notation. Option A is ∠BCA and Option C is ∠BAC.

2. Which angle below is acute, obtuse and a right angle?

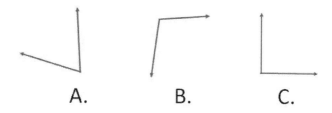

Answer: A. Acute, B. Obtuse, C. Right Angle

3. The four angles below all look congruent. But they are not all congruent, only two pairs of angles are congruent. Which two pairs of angles are congruent to each other?

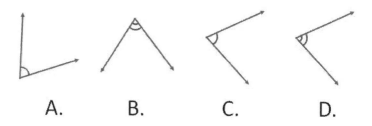

Answer: A and C are congruent, and B and D are congruent. This is identified by the arcs at the vertex of each angle.

10

4. Determine the value of angle x below:

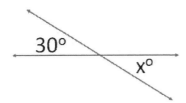

Answer: x = 30°, because opposite angles of intersecting lines are always equal.

5. Determine which pair of angles below are complementary and which pair are supplementary:

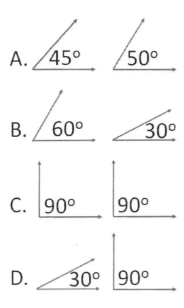

A. 45°   50°

B. 60°   30°

C. 90°   90°

D. 30°   90°

Answer: B. are complementary angles (add up to 90°), C. are supplementary angles (add up to 180°).

Chapter 4 Planar Lines

Lines can be identified as: coplanar, non-coplanar perpendicular and parallel. An example of some of these lines is shown in the 3-D diagram below (IJ intersects in the center of each horizontal square):

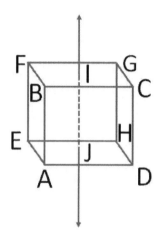

Coplanar (located in the same plane):

AB & CD

AB & EF

EF & GH

GH & CD

FB & GC

EA & HD

EF & IJ & CD

AB & IJ & GH

Do Not Intersect (lines that never intersect):

AB & CD

GC & AB

IJ & EA

EF & GC

IJ & AD

CD & EA

Perpendicular (lines at 90°) lines can be indicated by using ⊥:

AB & BC

CD & AD

AF & FG

HD & GH

Parallel (lines in the same plane that never intersect):

AB & IJ

EF & GH

EA & FB

Parallel Lines can be indicated by using: AB || IJ. Parallel Lines can also be indicated by using arrows in the lines:

If two parallel lines are both intersected by a third line, the following angles will then be equal:

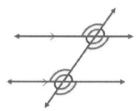

Therefore, if it is required to determine if two lines are parallel, one method to determine this is to investigate if an intersecting line results in congruent angles.

Practice Problems

1. Which pair of lines below are parallel?

A.

B.

C.

D.

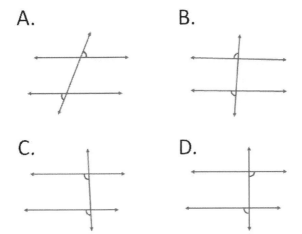

Answer: A and C are parallel lines. This is because their corresponding angles are congruent. The congruent angles for B and D are not corresponding angles.

Chapter 5 Triangles

A triangle is an object with three straight sides. Every triangle has three points, and therefore, three internal angles. The three internal angles of a triangle ALWAYS add up to 180°:

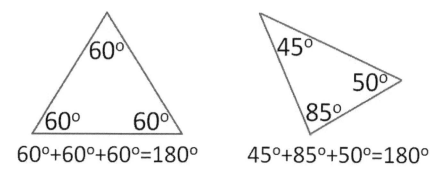

$$60°+60°+60°=180°\qquad 45°+85°+50°=180°$$

If two angles of a triangle are known, then the third angle can be calculated using the difference of the two angles from 180°.

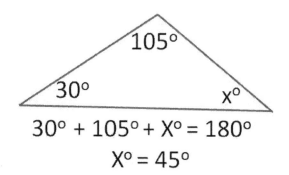

$$30° + 105° + X° = 180°$$
$$X° = 45°$$

The same method applies if one angle is known, and the other two angles are congruent:

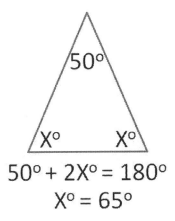

$$50° + 2X° = 180°$$
$$X° = 65°$$

There are three different types of triangles, scalene, isosceles and equilateral. For scalene triangles, all sides have a different length:

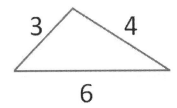

For isosceles triangles, two sides have the same length:

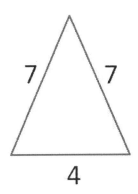

For equilateral triangles, all three sides are the same length (and in an equilateral triangle, all angles are 60°):

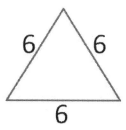

Triangles can also be referred to as acute, obtuse or a right triangle. In an acute triangle, all three angles are less than 90°:

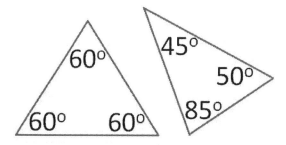

Note that one of the above acute triangles is also equilateral (all three sides same length), and the other acute triangle is scalene (all three sides different length). For right triangles, one angle is 90°:

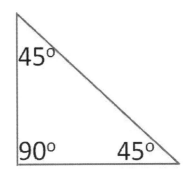

For an obtuse triangle, at least one angle is greater than 90°:

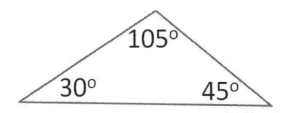

In a right triangle, the opposite side of the 90° angle is called the hypotenuse. And the 90° angle can be shown by using a small square box in the corner of the triangle:

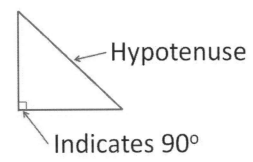

The angles of a triangle are referred to as the "interior angles." If a triangle is sitting on a flat surface, the sides make an angle with the surface. This is referred to as an "exterior angle":

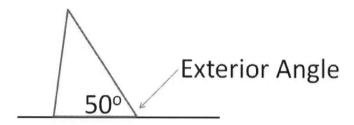

The interior angle and adjacent exterior angle ALWAYS add up to 180°:

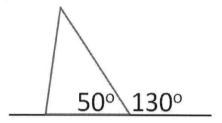

The altitude of a triangle is its height when it is sitting on a flat surface. Since a triangle has three sides, it can sit on a flat surface three different ways. So a triangle has three different altitudes.

For an acute triangle, the height is measured by drawing a perpendicular line from the surface to the opposite angle. In the below figure, an acute triangle is sitting on all three sides. The dotted line shows the height of the triangle on each side.

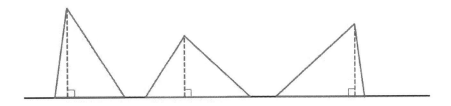

For a right triangle (a triangle with one ∠90°), the two sides are two of the heights of the triangle. And the third height is the distance from the hypotenuse to the ∠90°:

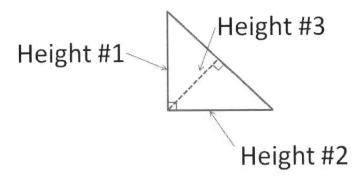

Height #1

Height #3

Height #2

For an obtuse triangle, two of the heights are on the exterior of the triangle, and the third height extends from the obtuse angle to the opposite side:

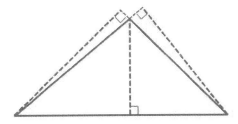

Congruent triangles are the same size and shape, and are identified using the ≅ symbol:

21

If all three sides of a triangle are equal, then the triangles are congruent. This is known as the side-side-side (SSS) rule:

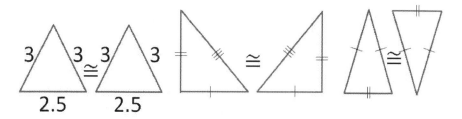

If two sides and a middle angle are equal to the same sides and middle angle of another triangle, then the triangles are congruent. This is known as the side-angle-side (SAS) rule:

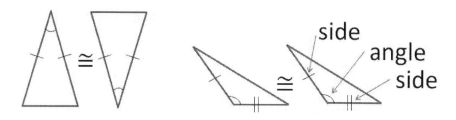

If two angles and their middle side are equal to the same two angles and side of another triangle, then the triangles are congruent. This is known as the angle-side-angle (ASA) rule:

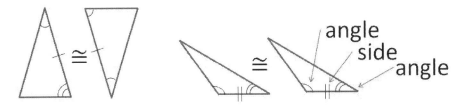

If two angles and their non-middle side are equal to the same two angles and non-middle side of another triangle, then the triangles are congruent. This is known as the angle-angle-side (AAS) rule:

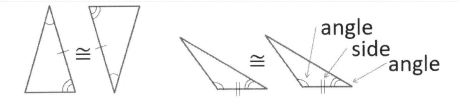

If the hypotenuse and leg of a right triangle are equal to the same hypotenuse and leg of another triangle, then the two triangles are congruent. This is known as the hypotenuse-leg (HL) rule:

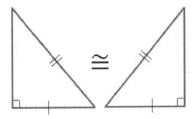

Triangles can be "named" by referring to the letter notated at each angle. When naming triangles, order matters. For example, we can refer to the triangle below as triangle ABC. However, that is not the same thing as referring to the triangle as BCA:

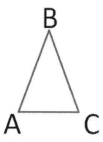

This is important when referring to triangles. For example below we can say that triangle ABC ≅ DEF. But we cannot say that BCA ≅ DEF. The corresponding angles must match:

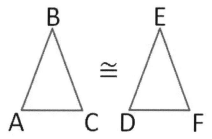

If two triangles are congruent, then all corresponding angles and sides are equal to each other.

The two smallest sides of a triangle combined together are always larger than the largest side of a triangle:

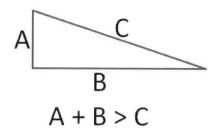

$$A + B > C$$

The length of each side of a triangle is always proportional to the size of the opposite angle:

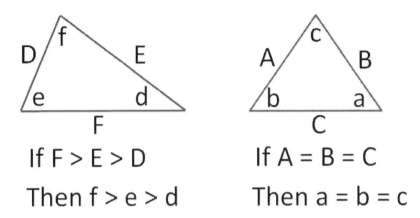

If F > E > D

Then f > e > d

If A = B = C

Then a = b = c

Practice Problems

1. What is the sum of the internal angles of a triangle?

Answer: 180°. All internal angles of a triangle add up to 180°.

2. Determine the value of x below.

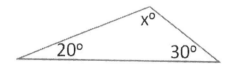

Answer: all the internal angles add up to 180°. We know two of the angles add up to 50° (20° + 30° = 50°), so the third angle must be 130° (50° + 130° = 180°).

3. Determine the value of x below.

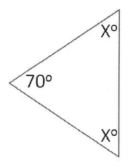

Answer: all the internal angles add up to 180°. We know one of the angles, and we know the other two angles are equal. Therefore:

$$70° + x° + x° = 180°$$

$$2x° = 110°$$

$$x° = 55°$$

4. Which triangle below is scalene, isosceles and equilateral?

A.   B.   C.

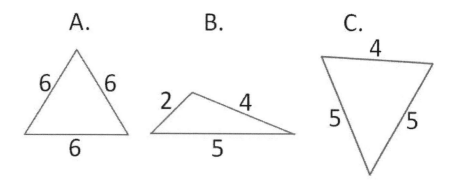

Answer: scalene B, isosceles C, equilateral A

5. Which triangle below is acute, obtuse and right?

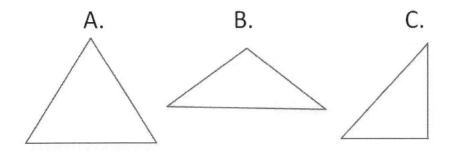

A.   B.   C.

Answer: acute A, obtuse b, right C

6. Which side is the hypotenuse of a right triangle?

Answer:

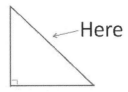

Here

7. Determine the value of the exterior angle x below.

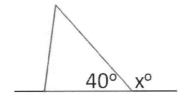

40° x°

Interior Angle + Exterior Angle = 180°

$40° + x° = 180°$

$X° = 140°$

8. What are the five rules that can be used to determine if two triangles are congruent?

Answer:

Side-Side-Side (if all three sides of each triangle are equal)

Angle-Side-Angle (if two angles and their middle side are equal)

Side-Angle-Side (if two sides and their middle angle are equal)

Angle-Angle-Side (if two angles and an adjacent side are equal)

Hypotenuse-Leg (if the hypotenuse and leg of a right triangle are equal)

9. True or False, A+B > C?

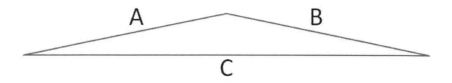

Answer: True, the two smallest sides of a triangle, when combined, are always larger than the largest side of a triangle.

Chapter 6 Polygons

Squares, triangles, rectangles and pentagons are all polygons. But circles are not polygons. Polygons are defined as closed shapes made of straight lines that do not intersect. The following are examples of polygons:

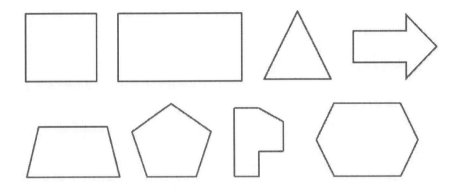

But lines, circles and open shapes are NOT polygons:

Polygons can be divided into convex polygons, and non-convex polygons. Non-convex polygons have interior points:

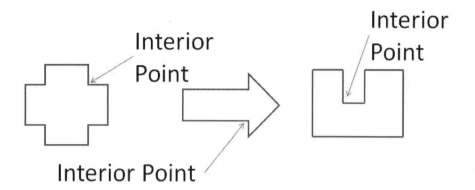

Convex polygons do not have interior points (ie squares, triangles, rectangles, etc).

Polygons can be notated by letters identified on each of their endpoints, for example:

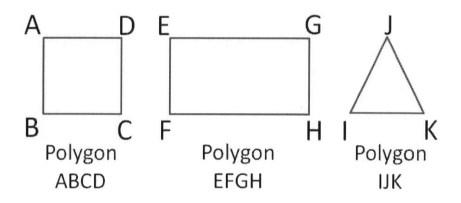

Convex polygons are categorized by the number of sides they have:

In a regular polygon, all sides are the same length and all internal angles are the same size:

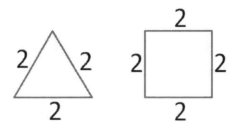

The interior angles of a triangle add up to 180°. To get the interior angles of a polygon, just divide the polygon into triangles:

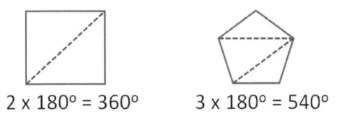

2 x 180° = 360°   3 x 180° = 540°

The internal angles of a square add up to 360°, and the internal angles of a pentagon add up to 540°. The interior angles of a polygon can be calculated by using the following equation:

Total interior angles of polygon = (n-2) x 180

Where *n* = the total # of sides of the polygon.

See below an example of this equation:

Triangle has three sides:
(3-2) x 180 = 180°

Quadrilateral has four sides:

(4-2) x 180 = 360°

Hexagon has six sides:
(6-2) x 180 = 720°

The internal angles for regular polygons are always equal, therefore the internal angle of a regular polygon can be calculated using the following equation:

Internal angle of a regular polygon = (n-1)180 / n

Where *n* = the total # of sides of the polygon.

See below an example of this equation:

Triangle has three sides:
(3-2)180 / 3 = 60°

Quadrilateral has four sides:
(4-2)180 / 4 = 90°

Hexagon has six sides:
(6-2)180 / 6 = 120°

The internal angle of a regular triangle is always 60°, the internal angle of a regular quadrilateral is always 90° and the internal angle of a regular hexagon is always 120°.

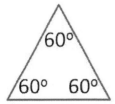 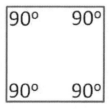 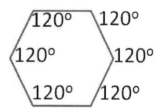

The exterior angles of a polygon ALWAYS add up to 360°:

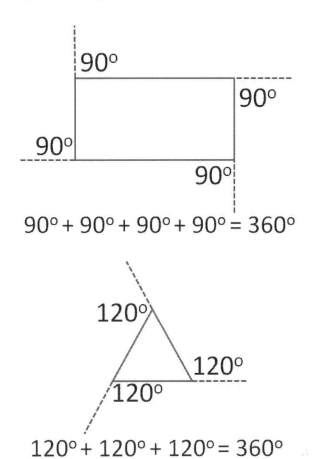

$$90° + 90° + 90° + 90° = 360°$$

$$120° + 120° + 120° = 360°$$

Practice Problems

1. Which of the below figures are polygons?

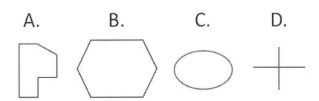

A.   B.   C.   D.

Answer: A and B are polygons. C is not a polygon because it does not have straight lines. D is not a polygon because it is not a closed object.

2. Which of the below figures are convex polygons?

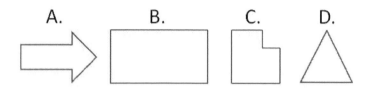

Answer: B and D are convex polygons because they do not have any interior points. A and C both have interior points, and therefore are not convex polygons.

3. Attach the shape name to each shape listed below:

Triangle, Quadrilateral, Pentagon, Hexagon, Octagon

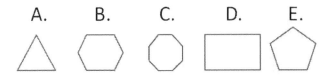

Answer: Triangle A, Quadrilateral D, Pentagon E, Hexagon B, Octogon C.

4. Which of the following are regular polygons?

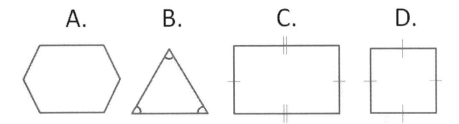

Answer: B and D are regular polygons. Since the angles of B are all equal, we know the sides are all equal length. Since the sides of D are all equal, we know the angles are all equal.

5. Determine the sum of all the interior angle degrees of an octagon:

Answer: An octagon has eight sides

$$(8-2) \times 180 = 1080°$$

6. Determine the sum of all the exterior angles of a square.

Answer: 360°, the sum of all exterior angles of any polygon is always 360°.

Chapter 7 Quadrilaterals

Quadrilaterals are polygons with four sides, these include squares and rectangles.

A parallelogram is a quadrilateral where the opposite sides are parallel and the opposite angles are equal. The following quadrilaterals are parallelograms:

These quadrilaterals are NOT parallelograms:

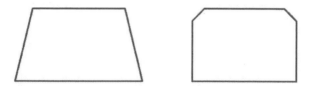

A rhombus is a quadrilateral with four congruent sides:

A square is technically a rhombus. But it is generally referred to only as a square.

The diagonals of a rhombus are always perpendicular:

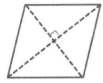

A trapezoid is a quadrilateral with only one pair of parallel sides (the non-parallel sides are called the legs):

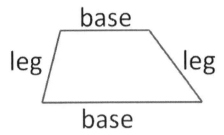

In an isosceles trapezoid, both legs are congruent:

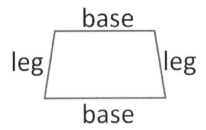

Practice Problems

1. Is a square a rhombus?

Answer: yes

2. Which of the shapes below are parallelograms?

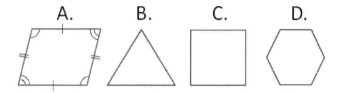

Answer: A and C. B and D are not a quadrilaterals and therefore cannot be a parallelograms.

3. True or False: The diagonals of a rhombus are perpendicular.

Answer: True

Chapter 8 Similarity

Two polygons are considered "similar" if both their corresponding angles are congruent AND their sides are proportional to each other. The sides do not have to be equal, only proportional. In order to be proportional, the ratio of corresponding sides must be equal for all sides:

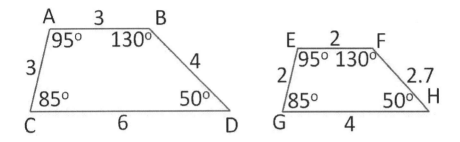

Here:

$$\frac{EF}{AB} = \frac{2}{3}$$

$$\frac{EG}{AC} = \frac{2}{3}$$

$$\frac{GH}{CD} = \frac{4}{6} = \frac{2}{3}$$

$$\frac{FH}{BD} = \frac{2.7}{4} = \frac{2}{3}$$

Since each corresponding side is equal to the same fraction, all the sides are proportional and the two polygons are similar. Since the fraction is 2/3, the scale factor is 2:3

If at least two corresponding angles for two triangles are equal, then the two triangles are similar:

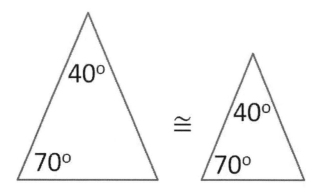

If two corresponding sides are proportional and their middle angle is equal, then the two triangles are similar:

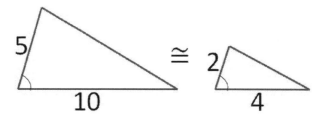

Since 2/5 = 4/10, the corresponding sides are proportional and the scale factor for these similar triangles is 2:5.

Practice Problems

1. Are the below shapes similar?

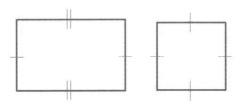

Answer: no because the sides are not proportional to each other.

2. Are the below shapes similar?

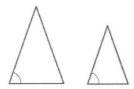

Answer: they appear to be similar, but it cannot be confirmed without knowing if at least two angles are the same or if the lengths of the sides are proportional. Therefore, it cannot be determined with the information provided.

3. Are the below shapes similar?

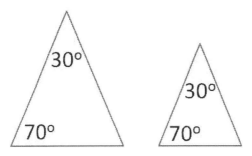

Answer: Yes, because we know two corresponding angles are equal, the third angles must also be equal and the shapes similar.

Chapter 9 Right Triangles

A right triangle is a triangle where one angle is 90°.

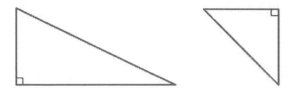

Pythagorean Theorem – in a right triangle, the square of the hypotenuse (the side opposite the 90° angle) equals the sum of the squared legs:

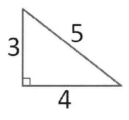

$$3^2 + 4^2 = 5^2$$

$$9 + 16 = 25$$

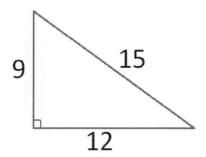

$$9^2 + 12^2 = 15^2$$

$$81 + 144 = 225$$

If the hypotenuse squared is larger than the sum of the legs squared, then the triangle is not a right triangle, it is an obtuse triangle. Similarly, if the hypotenuse squared is smaller than the sum of the legs squared, then the triangle is an acute triangle.

In 45-45-90 triangles (a term which refers to the size of the three angles), the legs are always equal, and the hypotenuse $= l\sqrt{2}$ as shown below:

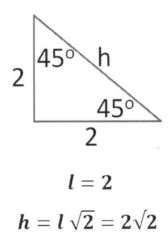

$$l = 2$$

$$h = l\sqrt{2} = 2\sqrt{2}$$

In 30-60-90 triangles (which also refers to the size of the three angles), there is a long leg and a short leg. where:

$$hypotenuse = 2 \times shortest\ leg$$

And:

$$longest\ leg = \sqrt{3} \times shortest\ leg$$

See an example below:

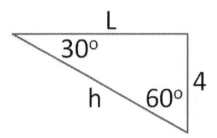

$$hypotenuse = 2 \times shortest\ leg = 2 \times 4 = 8$$

$$longest\ leg\ (L) = \sqrt{3} \times shortest\ leg = 4\sqrt{3}$$

Practice Problems

1. Determine the value of x below:

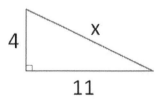

Answer: Use the Pythagorean Theorem to solve:

$$4^2 + 11^2 = x^2$$

$$137 = x^2$$

$$11.7 = x$$

2. Determine the value of x below:

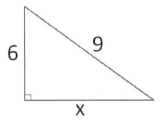

Answer: Use the Pythagorean Theorem to solve:

$$6^2 + x^2 = 9^2$$

$$x^2 = 81 - 36$$

$$x = 6.7$$

3. Determine the value of x below:

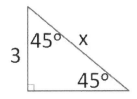

Answer: x is the hypotenuse of a 45-45-90 triangle, so we can use the equation $h = l\sqrt{2}$. (NOTE: only one leg is provided, but since this is a 45-45-90 triangle, both legs are always equal).

$$x = l\sqrt{2} = 3\sqrt{2} = 4.2$$

4. Determine the value of x below:

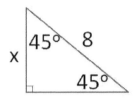

Answer: x is the leg of a right triangle. This is a 45-45-90 triangle, so we can use the equation $h = l\sqrt{2}$.

$$x = l\sqrt{2}$$

$$8 = x\sqrt{2}$$

$$x = 5.7$$

5. Determine the value of x below:

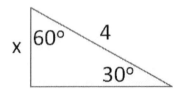

Answer: x is the leg of a 30-60-90 triangle. In a 30-60-90 triangle, the shortest leg is always half the length of the hypotenuse. The hypotenuse is 4, therefore x = 2.

6. Determine the value of x below:

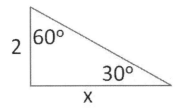

Answer: x is the leg of a 30-60-90 triangle. In a 30-60-90 triangle, the longest leg = $\sqrt{3} \times shortest\ leg$. Therefore:

$$x = 2\sqrt{3} = 3.5$$

Chapter 10 Trigonometry

Trigonometry is the use of special calculations which use the size of an angle to calculate the length of every side of a triangle. But this only works for right (90°) triangles.

The calculations involve three different functions known as cosine (cos), sine (sin) and tangent (tan). These functions are multipliers, which uses the value of one angle to calculate the length of one side.

The cosine (cos) of an angle is used to calculate the length of the adjacent side and the hypotenuse.

The sine (sin) of an angle is used to calculate the length of the opposite side and the hypotenuse.

The tangent (tan) of an angle is used to calculate the length of the opposite side and the adjacent side.

The figure below shows the opposite and adjacent sides from ∠c, as well as the hypotenuse:

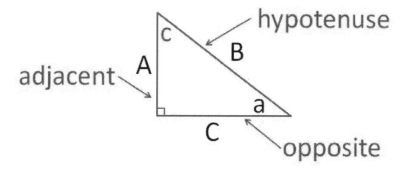

The next figure below shows the opposite and adjacent sides from ∠a, as well as the hypotenuse:

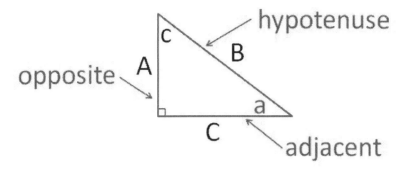

As you can see, the "opposite" side is the side not touching the angle. And the adjacent side is always the side that is touching the angle.

The equations for cos, sin and tan are as follows (where θ refers to angle size):

$$\cos\theta = \frac{adjacent\ side}{hypotenuse}$$

$$\sin\theta = \frac{opposite\ side}{hypotenuse}$$

$$\tan\theta = \frac{opposite\ side}{adjacent\ side}$$

An easy way to remember these equations is to remember the acronym phrase: SOH-CAH-TOA.

We can show this a different way using the below triangle:

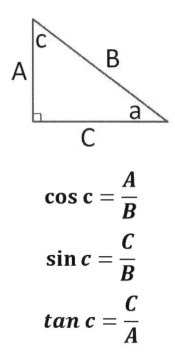

$$\cos c = \frac{A}{B}$$

$$\sin c = \frac{C}{B}$$

$$\tan c = \frac{C}{A}$$

Below is an example of one of these calculations where we try to find the lengths of the two unknown sides. First to calculate C:

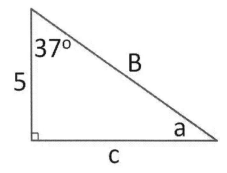

$$\tan 37 = \frac{opposite\ side}{adjacent\ side} = \frac{C}{5}$$

$$5\tan 37 = C$$

$$3.77 = C$$

And now to calculate B:

$$\cos 37 = \frac{adjacent\ side}{hypotenuse} = \frac{5}{B}$$

$$B = \frac{5}{\cos 37}$$

$$B = 6.26$$

Remember, sin, cos and tan functions can only be solved using a calculator. Below is an image of sin, cos and tan functions on a calculator:

Below is an example of what the above problem would look like on a calculator screen:

$$5/\cos(37)$$

After hitting "Enter" or "CE" the value on the calculator will show the answer as "6.26."

Sin, cos and tan functions can also be used to calculate the angle if the lengths are already known. The formulas are as follows (where θ refers to angle size):

$$cos^{-1} \times \frac{adjacent\ side}{hypotenuse} = \theta$$

$$sin^{-1} \times \frac{opposite\ side}{hypotenuse} = \theta$$

$$tan^{-1} \times \frac{opposite\ side}{adjacent\ side} = \theta$$

In the below figure, ∠c is calculated using three different methods (cos^{-1}, sin^{-1}, tan^{-1}):

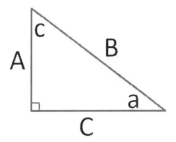

$$c° = cos^{-1} \times \frac{adjacent\ side}{hypotenuse} = cos^{-1} \times \frac{A}{B}$$

$$c° = sin^{-1} \times \frac{opposite\ side}{hypotenuse} = sin^{-1} \times \frac{C}{B}$$

$$c° = tan^{-1} \times \frac{opposite\ side}{adjacent\ side} = tan^{-1} \times \frac{C}{A}$$

Below is an example calculation for ∠c using all three methods (cos^{-1}, sin^{-1}, tan^{-1}):

53

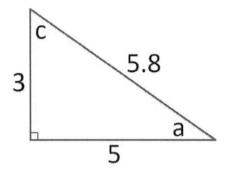

$$c° = cos^{-1} \times \frac{adjacent\ side}{hypotenuse} = cos^{-1} \times \frac{3}{5.8} = 58°$$

$$c° = sin^{-1} \times \frac{opposite\ side}{hypotenuse} = sin^{-1} \times \frac{5}{5.8} = 58°$$

$$c° = tan^{-1} \times \frac{opposite\ side}{adjacent\ side} = tan^{-1} \times \frac{5}{3} = 58°$$

Entered into a calculator, the tan equation above would look like this:

$$tan^{-1}(5/3)$$

The law of sines, as shown below, is true for any triangle:

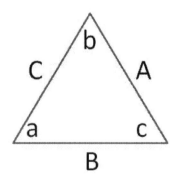

$$\frac{\sin a}{A} = \frac{\sin b}{B} = \frac{\sin c}{C}$$

See below an example problem using the law of sines:

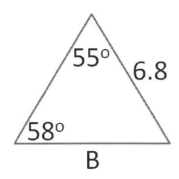

$$\frac{\sin 58°}{6.8} = \frac{\sin 55°}{B}$$

$$B = \frac{6.8 \sin 55°}{\sin 58°} = 6.6$$

The law of cosines, as shown below, is true for any triangle:

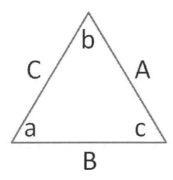

$$A^2 = B^2 + C^2 - 2BC(\cos a)$$
$$B^2 = A^2 + C^2 - 2AC(\cos b)$$
$$C^2 = A^2 + B^2 - 2AB(\cos c)$$

See below an example problem for the calculation of ∠c using the law of cosines:

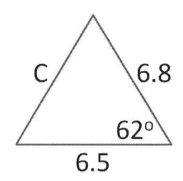

$$C^2 = A^2 + B^2 - 2AB(\cos c)$$
$$C^2 = 6.8^2 + 6.5^2 - 2(6.8 \times 6.5)\cos 62°$$
$$C^2 = 47.0$$
$$C = 6.9$$

Practice Problems

1. Determine the value of x below.

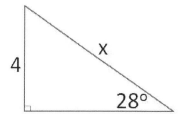

Answer: the opposite angle is known, and we are looking for the hypotenuse. Using SOH-CAH-TOA, we can see that sine applies to the opposite side and hypotenuse (S**OH**-CAH-TOA). Therefore we can use the sine function.

$$sin28° = \frac{opposite}{hypotenuse} = \frac{4}{x}$$

$$x = \frac{4}{sin28°} = 8.5$$

2. Determine the value of x below.

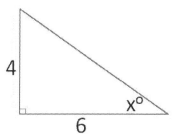

Answer: The two known sides are the sides opposite and adjacent to the unknown angle. Using SOH-CAH-TOA, we can see that tangent applies to the opposite side and adjacent (SOH-CAH-T**OA**). Therefore we can use the tangent function:

$$x = tan^{-1} \times \frac{opposite\ side}{adjacent\ side}$$

$$x = tan^{-1} \times \frac{4}{6}$$

$$x = 33.7°$$

3. Determine the value of x below.

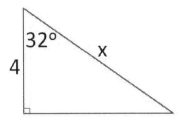

Answer: Using SOH-CAH-TOA, we can see that cosine applies to the adjacent side and hypotenuse (SOH-C**AH**-TOA). Therefore we can use the cosine function:

$$cos32° = \frac{adjacent}{hypotenuse} = \frac{4}{x}$$

$$x = \frac{4}{cos32°} = 4.7$$

4. Determine the value of x below, rounded to the nearest tenth:

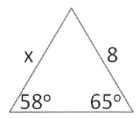

Answer: We can use the Law of Sines to calculate the value of x:

$$\frac{\sin 58°}{8} = \frac{\sin 65°}{x}$$

$$x = \frac{8 \sin 65°}{\sin 58°} = 8.6$$

5. Determine the value of x below.

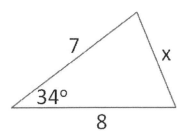

Answer: since the opposite angle is known, as well as the other two sides, we can use the Law of Cosines to solve.

$$x^2 = A^2 + B^2 - 2AB(\cos x)$$

$$x^2 = 7^2 + 8^2 - 2 * 7 * 8(\cos 34)$$

$$x = 4.5$$

Chapter 11 Circles

General definitions relating to circles are shown below:

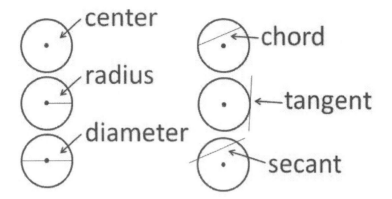

Congruent circles are circles with the same radius (or diameter):

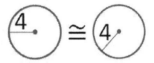

Concentric circles are circles with the same center:

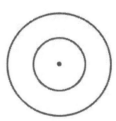

The following circles are tangent circles:

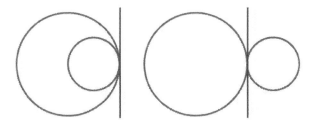

Tangents are always perpendicular to the radius of a circle:

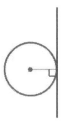

An inscribed angle, is an angle with the vertex on the circle. In the below figure, we can say the angle is inscribed in the circle, or the circle circumscribes the angle:

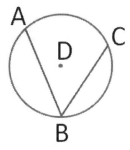

An arc is a section of a circle, which can be expressed in degrees just like an angle. The arc of a circumscribed angle is always twice as large as the angle:

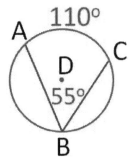

In the below figure, we can say the square circumscribes the circle, or the circle is inscribed by the square:

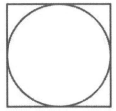

An arc of a circle has the same measurement as an angle inscribed in the circle, if the vertex is at the center:

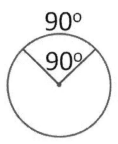

The degree measurement of an entire circle is 360°.

A semi-circle is exactly one-half of a full circle, and its degree measurement is 180°.

The arc length of an arc of a circle is calculated as follows:

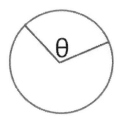

$$Arc\ Length = \frac{\theta}{360}(2\pi r)$$

For an example problem, given the below arc and chord, calculate the radius of the circle below:

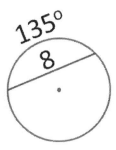

Step 1: draw two radii, one to each end of the 135° arc (the resulting angle will also be 135°):

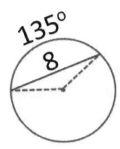

Step 2: divide the new triangle in half, and then use trigonometry to calculate the length of the radius (hypotenuse):

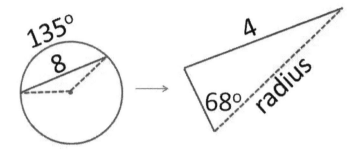

Step 3: the longer side is the *opposite* side from the 68° angle (we round up from 67.5°), and the radius is the hypotenuse. Since

$$\sin = \frac{opposite\ side}{hypotensue}$$

, the radius can be calculated as follows:

$$\sin 68° = \frac{4}{hypotenuse}$$

$$hypotenuse = \frac{4}{\sin 68°}$$

$$hypotenuse = radius = 4.3$$

When a chord and a tangent meet, they form both an angle and an arc. The arc will always be twice as large as the angle:

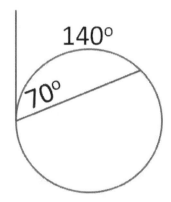

When an angle is not inscribed on a circle, the radius can still be calculated as the arc on the circle is known. Three examples of different intersection types are shown below:

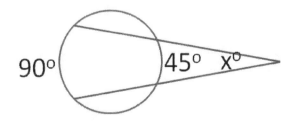

$$x° = \frac{90° - 45°}{2} = 22.5°$$

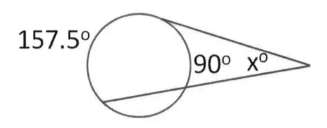

$$x° = \frac{157.5° - 90°}{2} = 33.75°$$

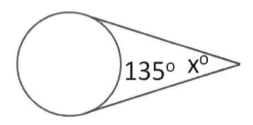

$$x° = \frac{225° - 135°}{2} = 45°$$

Two intersecting chords create two sets of line segments. If the lengths of each of the line segments are multiplied by each other, their products will always be equal:

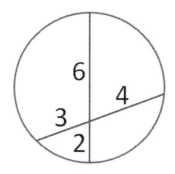

$$3 \times 4 = 6 \times 2 = 12$$

For two secants with the same common point, the following calculation is always true:

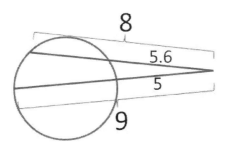

$$8 \times 5.6 = 9 \times 5 = 45$$

And for a secant and a tangent, the following calculation is always true:

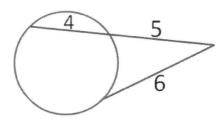

$$(4 + 5) \times 4 = 6^2$$

Practice Problems

1. Determine the angle of the arc below.

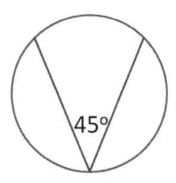

Answer: since the vertex of the angle is on the circle, the arc is twice the size as the angle of the vertex. Therefore the arc is 2 x 45° = 90°.

2. Determine the angle of the arc below.

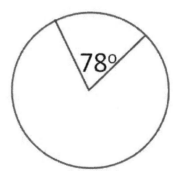

Answer: since the vertex of the angle is on the center of the circle, the arc is equal to the angle of the vertex. Therefore the arc is 78°.

3. Determine the length of the arc below.

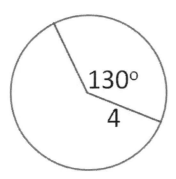

Answer: knowing the angle of the vertex and the radius of the circle, we can use the equation for arc length to solve.

$$Arc\ Length = \frac{\theta}{360}(2\pi r)$$

$$Arc\ Length = \frac{130°}{360}(2\pi 4)$$

$$Arc\ Length = 9.1$$

4. Determine the length of the arc below.

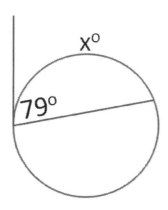

Answer: the angle formed by a chord and a tangent is half the size of an arc created by the same chord and tangent. Therefore, $x° = 2 \times 79° = 158°$.

Chapter 12 Perimeter and Area

The perimeter of an object is the length of all the sides added together. For a square, the perimeter is the length of one side x 4:

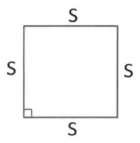

$$Square\ Perimeter = s + s + s + s = 4 \times s = 4s$$

The area of an object is the number of squares that can fit into the object. For most objects, this can be calculated by multiplying the length times the width:

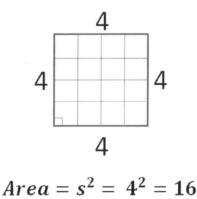

$$Area = s^2 = 4^2 = 16$$

The perimeter and area of a rectangle is calculated as follows below:

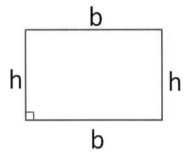

Rectangle Perimeter $= 2h + 2b$

Rectangle Area $= b \times h$

The object below is neither a square nor a rectangle:

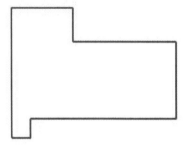

The perimeter and area of this irregular shape can be calculated by dividing it into known shapes, in this case two rectangles and one square.

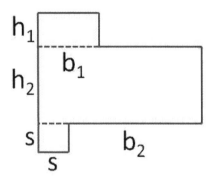

The perimeter of this shape is not the perimeter of all the shapes added together; instead it is the outside perimeter of all the shapes combined into one:

$$Perimeter = h_1 + b_1 + h_1 + (b_2 - b_1) + h_2 + (b_2 - s) + s + s + s + h_2$$

$$Area = h_1 \times b_1 + h_2 \times b_2 + s^2$$

The perimeter and area of a parallelogram is calculated as follows below:

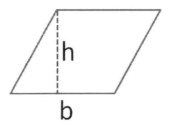

$$Parallelogram\ Perimeter = 2b + 2h$$

$$Parallelogram\ Area = b \times h$$

The area of a triangle is calculated as follows:

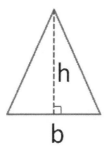

$$Triangle\ Area = \frac{1}{2}(b \times h)$$

The area of a right triangle is calculated as follows:

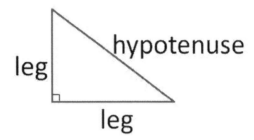

$$Right\ Triange\ Area = \frac{leg \times leg}{2}$$

The area of a rhombus (a parallelogram with equal sides) is calculated as follows:

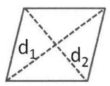

$$Area\ of\ Rhombus = \frac{d_1 d_2}{2}$$

The area of a trapezoid is calculated as follows:

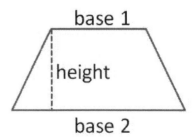

$$Area\ of\ Trapezoid = h\frac{b_1+b_2}{2}$$

The area and circumference of a circle is calculated as follows (circumference is the same thing as perimeter):

$$Area\ of\ Circle = \pi r^2$$

$$Circumference\ of\ Circle = 2\pi r$$

$$Circumference\ of\ Circle = \pi d$$

Practice Problems

1. Determine the perimeter and area of the rectangle below.

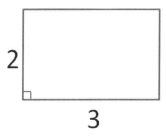

Answer:

$$Perimeter = 2l + 2w = 2 \times 3 + 2 \times 2$$

$$Perimeter = 10$$

$$Area = l \times w = 2 \times 3 = 6$$

2. Determine the perimeter and area of the irregular shape below.

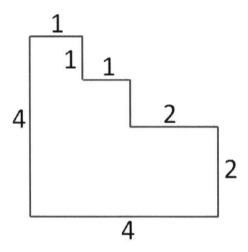

Answer: Break the irregular shape into known shapes, in this case rectangles and squares.

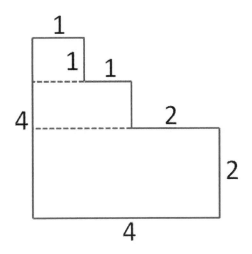

$$Perimeter = \ 4 + 1 + 1 + 1 + 1 + 2 + 2 + 4 = 16$$

$$Area\ of\ Small\ Square = 1 \times 1 = 1$$

$$Area\ of\ Small\ Rectangle = 2 \times 1 = 2$$

$$Area\ of\ Large\ Rectangle = 4 \times 2 = 8$$

$$\underline{Total\ Area = 1 + 2 + 8 = 11}$$

3. Determine the perimeter and area of the triangle below (HINT: Pythagorean Theorem is needed to get the height)

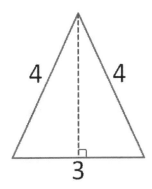

Answer: Perimeter is calculated by adding the sides and area is calculated after determining the height.

$$Perimeter = 3 + 4 + 4 = 11$$

Since we know this is an isosceles triangle, the height intersects the base at its midpoint. The triangle can be broken into two equal triangles to calculate the height.

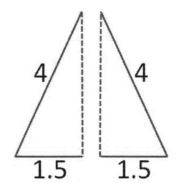

Using Pythagorean Theorem, the height of the triangle can be calculated:

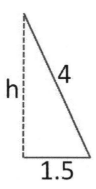

$$h^2 + 1.5^2 = 4^2$$

$$h^2 = 13.75$$

$$h = 3.7$$

Now that the height is known, the area of the triangle can be calculated:

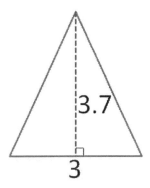

$$Area = \frac{1}{2} Base \times Height$$

$$Area = \frac{1}{2}(3 \times 3.7) = 5.6$$

4. Determine the area of the parallelogram below:

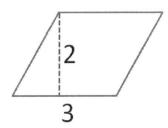

Answer:

$$Parallelogram\ Area = b \times h$$

$$Parallelogram\ Area = 2 \times 3 = 6$$

5. Determine the area of the rhombus below:

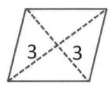

Answer:

$$Area\ of\ Rhombus = \frac{d_1 d_2}{2}$$

$$Area\ of\ Rhombus = \frac{3 \times 3}{2} = 4.5$$

6. Determine the area of the trapezoid below:

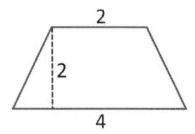

Answer:

$$Area\ of\ Trapezoid = h\left(\frac{b_1 + b_2}{2}\right)$$

$$Area\ of\ Trapezoid = 2\left(\frac{2 + 4}{2}\right) = 6$$

7. Determine the circumference and area of the circle below:

$$Circumference\ of\ Circle = 2\pi r = 4\pi$$

$$Circumference\ of\ Circle = 12.6$$

$$Area\ of\ Circle = \pi r^2 = \pi 2^2 = 12.6$$

Chapter 13 Volume and Surface Area

The surface area of an object is the total combined area of all the surfaces. The volume of an object is the total amount of space inside an object. It is measured as the total amount of cubes that can be put into the object.

A rectangular prism has six sides; four sides are faces, and two sides are bases. Since a rectangular prism has six surfaces, the surface area is the combined area of all six surfaces.

Surface area and volume of a rectangular prism are calculated as follows below:

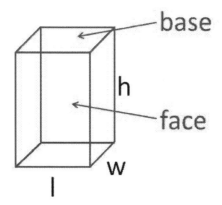

$$Surface\ Area = 2(wl + hl + wh)$$

$$Volume = l \times w \times h$$

A triangular prism has five sides; three sides are faces, and two sides are bases. The bases are triangles and the sides are quadrilaterals.

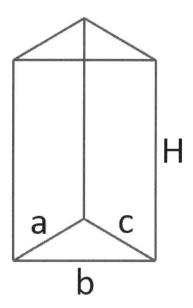

To calculate the surface area of a triangular prism, the height the triangle must be calculated. Remember from Chapter 5 Triangles that the height of a triangle must meet the base at a 90 degree angle. We can pick any base side we want, here we will use side b as the base:

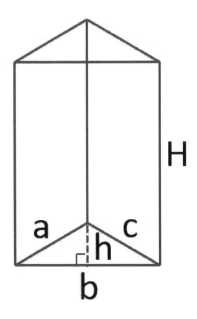

$$Surface\ Area\ = bh + (a + b + c)H$$

$$Volume = \frac{1}{4}H\sqrt{-a^4 + 2(ab)^2 + 2(ac)^2 - b^4 + 2(bc)^2 - c^4}$$

Volume and surface area of a regular pyramid are calculated as follows below (for a regular pyramid, we assume l=w):

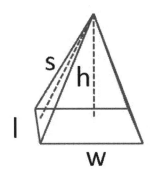

$$Surface\ Area\ of\ Pyramid = \frac{1}{2}l \times w \times s + l \times w$$

$$Volume\ of\ Pyramid = \frac{1}{3}l \times w \times h$$

Volume and surface area of a cylinder are calculated as follows below:

$$Surface\ Area = 2\pi rh + 2\pi r^2$$

$$Volume = \pi r^2 h$$

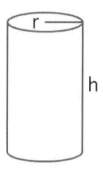

Volume and surface area of a cone are calculated as follows below:

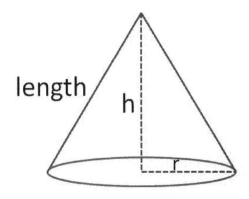

$$Surface\ Area\ of\ a\ Cone = \pi r l + \pi r^2$$

$$Volume\ of\ a\ Cone = \frac{1}{3}\pi r^2 h$$

Volume and surface area of a sphere are calculated as follows below:

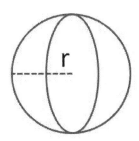

$$Surface\ Area\ of\ Sphere = 4\pi r^2$$

$$Volume\ of\ Sphere = \frac{4}{3}\pi r^3$$

Practice Problems

1. Determine the surface area and volume of the rectangular prism below.

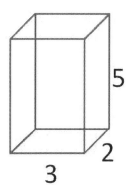

Answer:

$$Surface\ Area = 2(wl + hl + lw)$$

$$Surface\ Area = 2(2 \times 5 + 5 \times 3 + 3 \times 2)$$

$$Surface\ Area = 62$$

$$Volume = l \times w \times h$$

$$Volume = 3 \times 2 \times 5 = 30$$

2. Determine the surface area and volume of the triangular prism below.

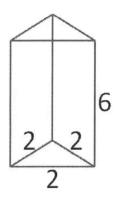

Answer:

$$Surface\ Area\ =\ bh + (a + b + c)H$$

The base (b) = 2, but we need to know the height (h). This is an equilateral triangle:

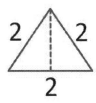

The height of the triangle can be found by using Pythagorean Theorem:

$$h^2 + 1^2 = 2^2$$
$$h^2 = 3$$
$$h = 1.7$$

Now that we know the height, we can calculate the area of the triangle:

$$Surface\ Area\ =\ bh + (a + b + c)H$$

$$Surface\ Area\ =\ 2 \times 1.7 + (2 + 2 + 2)6$$

$$Surface\ Area\ =\ 39.4$$

We can calculate the volume using the equation below:

$$Volume = \frac{1}{4}H\sqrt{-a^4 + 2(ab)^2 + 2(ac)^2 - b^4 + 2(bc)^2 - c^4}$$

$Volume$

$$= \frac{1}{4} \times 6\sqrt{-2^4 + 2(2 \times 2)^2 + 2(2 \times 2)^2 - 2^4 + 2(2 \times 2)^2 - 2^4}$$

$$Volume = 10.4$$

3. Determine the surface area and volume of the pyramid below with a slant height (s) of 5.4:

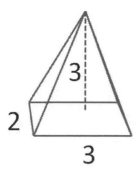

Answer:

$$Surface\ Area\ of\ Pyramid = \frac{1}{2}l \times w \times s + l \times w$$

$$Surface\ Area = \frac{1}{2}(2 \times 3 \times 5.4) + 2 \times 3$$

$$Surface\ Area = 22.2$$

$$Volume\ of\ Pyramid = \frac{1}{3}(l \times w \times h)$$

$$Volume\ of\ Pyramid = \frac{1}{3}(3 \times 2 \times 3)$$

$$Volume\ of\ Pyramid = 6$$

4. Determine the surface area and volume of the cylinder below (rounded to the nearest hundredth).

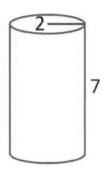

Answer:

$$Surface\ Area = 2\pi rh + 2\pi r^2$$
$$Surface\ Area = 2\pi 2 * 7 + 2\pi 2^2$$
$$Surface\ Area = 113$$

$$Volume = \pi r^2 h$$
$$Volume = \pi 2^2 \times 7$$
$$Volume = \pi r^2 h$$
$$Volume = \pi 2^2 \times 7 = 87.97$$

5. Determine the surface area and volume of the cone below.

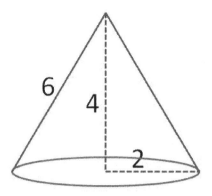

Answer:

$$Surface\ Area = \pi rl + \pi r^2$$
$$Surface\ Area = \pi 2 * 6 + \pi 2^2$$
$$Surface\ Area = 50.2$$

$$Volume = \frac{1}{3}\pi r^2 h$$
$$Volume = \frac{1}{3}\pi 2^2 * 4$$
$$Volume = 16.8$$

6. Determine the surface area and volume of the sphere below.

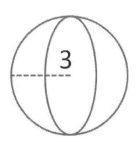

Answer:

$$\textbf{Surface Area} = \textbf{4}\boldsymbol{\pi}\textbf{r}^2 = \textbf{4}\boldsymbol{\pi}\textbf{3}^2 = \textbf{113}$$

$$Volume = \frac{4}{3}\pi r^3$$
$$Volume = \frac{4}{3}\pi 3^3 = 113$$

Chapter 14 Transformations

A transformation can be a translation, reflection or rotation. Below is a translation:

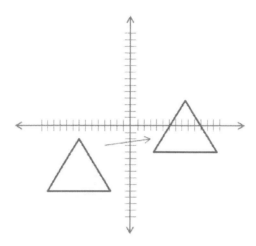

Here, the object has moved 16 spaces to the right (in the positive x-direction), and 7 spaces upward (in the positive y-direction). So we can say that the object has been translated as follows:

$$(x, y) -> (x + 16, y + 7)$$

Reflection is to reflect an object across the x or y-axis. Below is a reflection across the x-axis. The object is a mirror reflection on each side of the axis, the same distance apart from the axis on both sides:

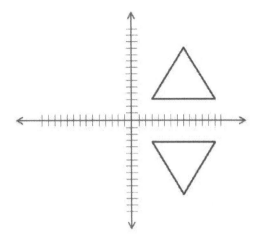

Below is a reflection across the y-axis. The object is a mirror reflection on each side of the axis, but this time the y-axis instead of the x-axis:

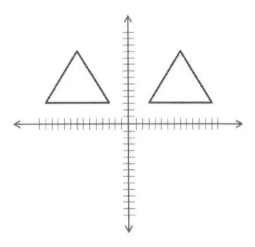

A rotation is to rotate an object around an axis. To rotate an object, we connect 2-3 lines from the origin (0, 0) to points on the object. Then we rotate the lines around the origin to the desired number of degrees. Below is the figure we want to rotate, say 180°:

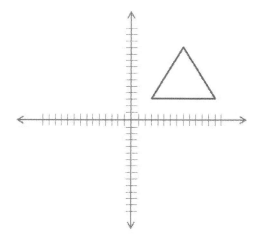

The lines are now drawn:

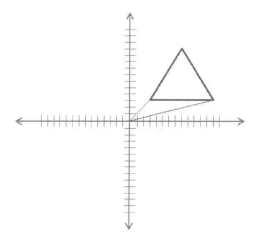

And the last step is to rotate the lines 180°:

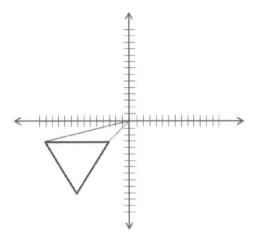

Practice Problems

1. Transform the figure below x-10, y+2:

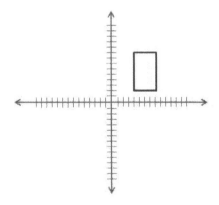

Answer:

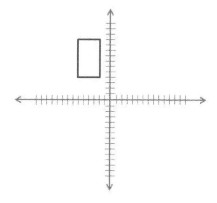

2. Reflect the below figure across the y-axis:

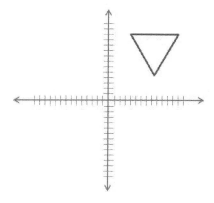

Answer:

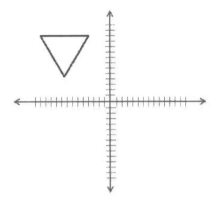

3. Rotate the below figure 270º:

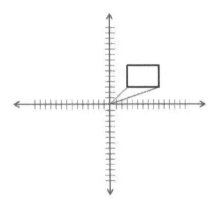

Answer:

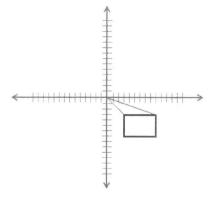

Questions or Comments?

This concludes Concise Geometry, we hope you enjoyed this book and hope you were able to go through the material quickly and efficiently. After completing this book, you should be ready to tackle more difficult Geometry problems, and you should be able to learn more in depth Geometry subjects fairly quickly.

This book is entirely self-published and has sold over 5,000 copies as of Spring, 2020.

If you have questions or comments, you can reach me at josiahcoates@gmail.com.

Below are my other two books:

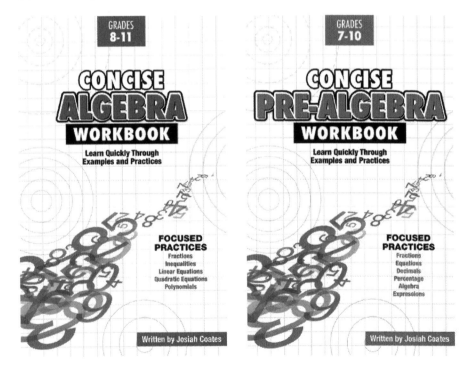

Made in the USA
Las Vegas, NV
28 November 2020